D0518157

CONTENTS

STORIES BY JAMES HENDERSON
ILLUSTRATED BY JOHN HARROLD
STORY COLOURING BY DORIS CAMPBELL

This book belongs to:

oren

RUPERT

ISBN 0-85079-206-1

John Harrold.

No 55   Printed in Great Britain. Canadian Distributors: Copp Clark Pitman Ltd., 2775 Matheson Blvd East, Mississauga, Ontario, Canada, L4W 4P7 Tel: (416) 238 6074 $9·95

RUPERT

*Rupert is going off to spend
The weekend with Pong-Ping, his friend.*

Rupert is off to spend the weekend with his chum, the Peke Pong-Ping whose house lies on the other side of Nutwood village. The quickest way there is through the grounds of Nutwood Court. The grounds are private but the house has stood empty so long that Rupert and his pals have got used to treating them as a short cut and so, as he's done so often, Rupert climbs the stile into the overgrown grounds.

and Little Yum

To get to Pong-Ping's Rupert goes
Along a short cut that he knows.

Then suddenly – oh, what a fright!
A burly stranger holds him tight!

Rupert's mind is so much on the weekend ahead that he almost jumps out of his skin when a heavy hand falls on his shoulder and a rough voice cries "Gotcha!" The hand belongs to a large unpleasant man. "Can't you read?" he growls, pointing back at the "Private" sign. "Oh, please, I didn't know . . . I thought . . ." Rupert stammers. "Enough!" the man cries. "You're trespassing, that's what. You're coming with me!"

He says that Rupert ought to know
It's private land, the sign says so!

RUPERT MEETS SIR JASPER

Then marching Rupert at his side
Towards the house he starts to stride.

Inside the house he tells him how
He's going to meet the master now.

The owner of the house looks grim
As Rupert is lead in to him.

"Please," Rupert quails, "I didn't know
You lived here now. Please let me go!"

"Come with you?" cries Rupert. "Why? Where?" But the man's only answer is to grab Rupert's wrist and march him towards the house. "You can't do this!" Rupert protests. "Oh, yes I can!" the man growls. "This is private property. You've no right here and now you're going to see my master. He'll decide what's to be done with you." With that he propels Rupert up the steps and into the house where he halts at a door and knocks on it.

"Come in!" snaps a voice. The man removes his cap before opening the door and pushing Rupert ahead of him. A man is standing at the fireplace, more smartly dressed than Rupert's captor but just as unpleasant looking. "Caught this one in the grounds, sir," the big man says. "I meant no harm," Rupert begins. "I didn't know anyone was living here." "Quiet!" the slim man barks. "You were trespassing in my grounds and I simply will not put up with that!"

RUPERT HEARS A WAIL

"Silence!" he barks; "I own these grounds.
The path you took is out of bounds!"

"I'll let you go, but make sure you
Tell all your friends to keep out too!"

As Rupert leaves, a doleful cry
Comes from a basement that's nearby.

When Rupert asks him, "What was that?"
Scrogg says it must have been a cat.

Rupert tries not to look as scared as he feels. "Oh, please," he begs. "Let me go. I'd no idea you'd moved into the house. It's been empty so long." "Quiet!" the man snaps again. He makes a show of thinking then he says, "I shall let you go just this once. But in return you shall tell all your friends they must stay off my land if they know what's good for them." Rupert nods eagerly. "Then Scrogg here will show you off the property," the man dismisses him.

Scrogg who plainly is some sort of servant of Nutwood Court's new owner marches Rupert out of the house. "You're lucky," he growls. "Sir Jasper ain't usually so gentle with trespassers. So you just do what he said." Just then as they pass a small dark basement window Rupert hears a wail. "What's that?" he asks. "It came from there." He points to the basement. "Nothing to worry you," snaps Scrogg. "Cats most like." And he bundles Rupert roughly away from the spot.

RUPERT'S CHUM HAS CALLERS

But Rupert can't believe his tale:
"It sounded such a mournful wail!"

"Forget it!" Scrogg warns angrily.
"It's safer just to let things be!"

At Pong-Ping's Rupert finds a pair
Of Oriental strangers there.

And as they leave the two men stare
Intently at the little bear.

Even as he's being marched back to the stile where he came in Rupert can't free his mind of that wail from the basement. "Like someone lost or lonely," he thinks. "Certainly wasn't a cat." As Scrogg sees him over the stile, with a warning to stay on the other side of it from now on, Rupert says what he's been thinking. Scrogg's face darkens. "Forget it. You'll find it safer to let things be," he snarls. Rupert shivers as he sets off for Pong-Ping's house.

As he hurries along, he wonders what Pong-Ping will make of all this – the man Scrogg, Sir Jasper and that doleful wail from the basement. There's something very odd going on there. "And something strange here as well," he murmurs as he turns into Pong-Ping's garden. His pal is seeing off two Oriental-looking men. Their plainly new clothes look like an attempt to make them less foreign-seeming. They stare at Rupert as they pass him.

RUPERT TELLS ABOUT THE WAIL

It seems they've come to ask the Peke
About a stranger whom they seek.

"A stranger!" Rupert cries, "I say,
But I've just met two on my way!"

Pong-Ping's intrigued and asks his friend
To tell his tale from start to end.

"Tomorrow," Pong-Ping says, "We'll go
To tell our policeman all we know!"

"I say, who are they?" whispers Rupert as the two Orientals disappear. "I'm not quite clear," the Peke replies. "They are from the Far High Mountains of my country. They speak almost no English and so someone in Nutwood sent them to me because, apart from our pal Tigerlily and her father the Conjuror who are away at the moment, I'm the only Chinese-speaker around here. They asked if any stranger has come to live here recently." "I've just met two," Rupert says.

Now, if there's one thing Pong-Ping likes it's a mystery. And this looks like one. A pair of Orientals turn up asking about newcomers to Nutwood and now Rupert says he knows of two. So as soon as Rupert is settled with a glass of milk the Peke demands his story. "You're sure it wasn't a cat you heard?" he asks when Rupert is done. "Certain," he's told. Later as he is showing Rupert where he's to sleep he says, "Something's wrong here. We must tell the police tomorrow."

RUPERT TELLS GROWLER IN VAIN

But Growler's upset when Pong-Ping
Tells him to look into this thing.

"That's private land!" he tells the pair.
"A well-known hunter now lives there."

Pong-Ping's still sure something's amiss
And says they must look into this.

As Nutwood Court comes into sight
Pong-Ping vows, "We'll go there tonight!"

Next morning Rupert and Pong-Ping hurry into Nutwood and pour out to PC Growler, the village bobby, the story of the wailing Rupert heard at Nutwood Court. "So we think you should look into it," Pong-Ping adds. "Oh, you do?" snaps Growler who dislikes being told what to do. "Well, let me tell you," he says, rising and ushering the two outside, "Sir Jasper is a highly respected trapper of animals for zoos. So be off with you. And stay off private property!"

"Oh, dear, we must have made a mistake," Rupert says as he and Pong-Ping walk away from the police station. "Mistake nothing!" snorts the Peke. "Just because that man at Nutwood Court is a sir and 'highly respected' Growler thinks there can't be anything wrong. Well, *we* think there is, so if Growler won't investigate we shall. We'll go there after dark tonight." Just then they pass Nutwood Court looming over the trees. "Oh, do you really think we should?" Rupert quavers.

10

RUPERT INVESTIGATES

Rupert reminds him of the way
They've twice been warned to stay away.

Rupert's reluctant but agrees.
Pong-Ping is set on this, he sees.

When darkness falls the pair set out
No one appears to be about.

Just as they reach the basement though
They see a light shine from below.

Rupert keeps fretting about Pong-Ping's plan to investigate Nutwood Court that night. He's far from happy about it and as they reach the Peke's house he says so: "After all, Sir Jasper and PC Growler have warned us about trespassing. I think we should leave well alone." "'Well' is what we wouldn't be leaving alone!" Pong-Ping snaps as they go indoors. "We both think something's wrong there. If you won't come I'll go alone." "Oh," sighs Rupert. "I suppose . . ."

And so that night our pair set out for Nutwood Court. Even the Peke whose idea this is, isn't so keen now that the time has come. But with Pong-Ping carrying a lantern and Rupert leading the way, they steal into the grounds of the old house. The night is light enough for them to do without the lantern as they make their way to the basement where Rupert heard the wailing. They are almost there when they stop in their tracks. There is a glow of light from the basement steps.

RUPERT HEARS THE WAIL AGAIN

"Scrogg!" whispers Rupert, but just then
The wailing sound starts up again.

"Quiet!" shouts Scrogg. He whacks the door
Then goes back to the house once more.

Pong-Ping now feels a deal less bold.
His fiery temper's gone quite cold!

But now it's Rupert's turn to say
They'll help the creature, come what may.

The pals creep forward to the shelter of a bush and from there they can see that the glow is from a lamp held by the servant Scrogg. Then they freeze as a heart-broken wail rises from inside the basement. "Quiet!" Scrogg snarls. "And stay quiet if you know what's good for you." With that he whacks the door with his stick, stumps up the basement steps and goes back to the house. The front door slams and then there is silence. Rupert and Pong-Ping exchange looks. Rupert can see that the wailing from the basement and the sight of the unpleasant Scrogg have had their effect on Pong-Ping who now looks a great deal less brave. "D – do you think we should go back?" he quavers. Rupert is just as scared but he shakes his head. "We can't," he hisses. "Not after that wail." Pong-Ping gulps and murmurs, "You're right." Just then the last light in the house goes out. "Now," Rupert says and they dash for the basement steps.

"Who's in there?" Rupert risks a call.
No word comes back – a gulp, that's all.

Then, pushing the door open wide
They gasp at what they find inside!

The creature blinking in the light
Is such an unexpected sight!

"A baby Yeti!" Pong-Ping cries.
He's not sure he believes his eyes.

Rupert and Pong-Ping reach the basement steps safely. They hold their breath. Will there be a sudden angry shout, a blaze of light from the house? No. Nutwood Court stays silent and dark. So they tiptoe down the steps. The door into the basement is a heavy affair with a bolt on the outside – luckily without a padlock. They listen. "Who's in there?" Rupert whispers as loudly as he dare. The answer is a gulp. The Peke switches on the lantern. They open the door . . .

Rupert and Pong-Ping stand speechless. The beam of their lantern reveals, cowering against the far wall, a creature such as Rupert has never imagined still less seen. It's smaller than he is and covered in fur from the top of its pointy head to the toes peeping from the bottom of dumpy little legs. Its eyes are big and startled as it tries to make out the figures behind the light. "I can't believe this," Pong-Ping breathes. "It's a – a baby Yeti, of all things!"

RUPERT HELPS FREE THE YETI

"What's that?" asks Rupert and hears tell
Of mountains where the Yeti dwell.

Pong-Ping explains they're very rare
Then soothes the little thing, "There, there!"

The Yeti seems to understand
As Pong-Ping leads it by the hand.

Then suddenly the two pals freeze,
The baby Yeti starts to sneeze!

"A Yeti?" repeats Rupert. The Peke nods. "Yeti live in the highest mountains of my country," he says. "So few outsiders have seen one that many don't believe in them. They grow big and look fearsome. In fact, they're very shy and are highly honoured by our mountain folk who'd do anything to protect them." He sighs: "But to some awful people a baby one like this would be worth a fortune." The little creature sobs. Pong-Ping pats its head and says what sounds like the Chinese for "There, there". The baby Yeti sniffs hard and stops sobbing. "We must get it out of here," urges Rupert. "Right," says Pong-Ping and takes the Yeti's hand. He says something in Chinese. "I said it will be safe with us and must be very quiet," he explains. "But I don't think it understood." Still, the Yeti is as quiet as the others as they start up the steps. Quiet, that is, until it sneezes. And how it sneezes! "Oh, no!" gasp Rupert and Pong-Ping.

RUPERT HAS TO RUN FOR IT

They start to bolt but don't get far
Before a torch shows where they are.

"Quick Scrogg!" they hear Sir Jasper say,
"Don't let the Yeti get away!"

At this they start to run, our three.
"This way!" calls Rupert. "Follow me!"

"Quick!" Rupert hisses. "Over here!"
As Jasper and his man draw near.

For a moment after the little Yeti's sneezing there is an awful silence. Rupert and the others hold their breath. Did anyone in the house hear it? They are beginning to think they may have been lucky when a window shoots up and they find themselves pinned in the beam of a powerful torch. Then Sir Jasper's voice rings out: "I see them! After them, Scrogg!" Rupert and Pong-Ping exchange horrified looks. The baby Yeti looks pleadingly up at them.

Lights come on in Nutwood Court and there is a clatter of feet as the two men race for the front door. "Quick!" cries Rupert. "To the stile!" At Rupert's cry Pong-Ping springs into life and, clasping the baby Yeti's hands, they make a run for it. The Yeti's little legs are fairly whirring as it's hustled along by the pals. But, fast as they are moving, they hear their pursuers getting closer. "Over here!" hisses Rupert and drags the others into the bushes.

RUPERT FINDS A HIDING PLACE

"Down!" Rupert orders, "Not a sound!"
And hopes so hard they won't be found.

They hold their breath and lie low while
The baddies hurry to the stile.

Says Rupert now they've lost those two
Make for the old gate's what they'll do.

The pals squeeze through but suddenly
Gasp in alarm at what they see . . .

"Where are you taking us?" pants Pong-Ping. But Rupert only hisses, "Sh! Get down both of you!" and pulls the pair of them down behind a clump of bushes. "And keep quiet!" At that moment the baby Yeti takes a deep breath as if it might sneeze again. Pong-Ping clamps a hand over its mouth and makes urgent Chinese sounds. It nods as if it understands and the Peke takes his hand away. A moment later Sir Jasper and Scrogg dash past, heading for the stile.

Rupert waits until the men are well past then whispers to Pong-Ping, "Now we've got to reach the old gate before they find they've been fooled." The old gate is the main entrance to the grounds and is where Rupert was headed when Scrogg stopped him. Pong-Ping makes comforting sounds to the baby Yeti then he and Rupert take its hands and the three make a dash for the gate. They reach it, squeeze through – and cry out at the sight that greets them.

16

RUPERT LEARNS ABOUT YUM

The men who came to see Pong-Ping!
But now they're stern and menacing . . .

Without a word they grab our two –
Whatever can they mean to do?

At this the Yeti starts to yell,
Though what it says the pals can't tell.

The men apologise and bow.
They know the pals are Yum's friends now!

Looming in the light of Pong-Ping's lamp are the two Orientals who came to his house asking about newcomers to Nutwood. "They must have heard about Sir Jasper being new," thinks Rupert. "That'll be why they're here." The strangers stare bleakly at the pals. One of them hisses angrily at Pong-Ping in Chinese. But before the Peke can reply he is grabbed by the man and swung off his feet. The other man lunges for Rupert and gathers him up struggling.

Suddenly the baby Yeti lets loose a torrent of high-pitched chatter. At once the pals are put down – respectfully. The men bow and one of them addresses Pong-Ping. The Peke translates: "They have come from the Far High Mountains to rescue the Yeti – his name's Yum – who was stolen from there. They traced him to Nutwood and when they saw us with him they thought we were the ones who took him. But now Yum has told them what really happened."

RUPERT HAS A BRIGHT IDEA

They tell Pong-Ping, he tells his chum,
They're here to save the Yeti, Yum.

Then Rupert says they must decide
Upon a place for Yum to hide.

Pong-Ping's house is where Yum will stay
Until it's time to get away.

From there the way to China's swift,
By way of Pong-Ping's direct lift.

Rupert looks on bewildered as Yum and one of the Orientals – he is Wing, his companion is Wang – chatter in Yeti talk. Wang translates into Chinese for Pong-Ping who tells Rupert: "Wing and Wang want to find Sir Jasper and punish him and Yum wants to know what we think." "It's more important to get Yum away from here at once," says Rupert. "The best place is your house." "You're right," the Peke agrees. And just then Yum begins to yawn and yawn.

When Yum has been told what the pals think and has agreed, the little party sets out for Pong-Ping's house with Wing carrying the little Yeti who is soon fast asleep. As they go Rupert wonders how the three can be got safely away from Nutwood and Sir Jasper. Then – got it! Pong-Ping's lift! In case you don't know Nutwood well, it should be explained that the Peke has a lift in his garden that goes right through the world, all the way to the part of China he comes from.

18

Pong-Ping says "Yes, that is the way
To get Yum home without delay."

The men agree but tell them how
There's no way they can wake Yum now.

They'll keep a watch throughout the night
While Little Yum is sleeping tight.

"Before I turn in, one last thing,
I'll check the lift," declares Pong-Ping.

"Just what I was thinking!" cries Pong-Ping when Rupert mentions his idea of using the lift to get Yum and the others away. But there's to be no putting the plan to work this night. For though Wing and Wang think the plan splendid – despite finding it hard to believe in the lift – they say firmly that Yum must be asked if he agrees. "Then wake him and ask," urges the Peke. "Impossible!" he's told. Baby Yeti, it seems, must be left to waken when they are ready.

There's nothing for it, then, but to wait for Yum to waken in his own time. Wing and Wang say they're sorry, but it's out of the question to take a so-honoured Yeti on any journey, particularly one as strange as a lift journey to China, without its approval. So with Yum tucked up on a couch and Wing and Wang keeping watch, the pals decide to get some sleep too. "Go on up," Pong-Ping tells Rupert. "I'll make sure the lift's working before I turn in."

RUPERT IS ACCUSED

Next day Pong-Ping lets Rupert know
That Yum's awake – it's time to go!

Soon Rupert's ready but before
They leave someone knocks at the door.

With Growler stands Sir Jasper who
Cries, "They took my wild beast, those two!"

But PC Growler's mystified
And says he'll take a look inside.

It is morning when Rupert is wakened by Pong-Ping. The Peke has been up some time and is full of his news. "Little Yum has agreed to go home by my lift," he announces. "Wing and Wang asked him as soon as he woke and he said, yes." "Then let's get moving now!" Rupert says. "I'll get dressed and join you downstairs." When he goes down the others are waiting for him. As he arrives there is a knock on the front door. Being nearest, Rupert opens it . . .

"They're the ones who took my valuable wild beast, constable!" Sir Jasper's angry words and accusing finger greet Rupert. PC Growler looks stern between Sir Jasper and Scrogg who is carrying a net. "Make him hand it over!" Sir Jasper cries. "Yum isn't a wild beast!" Rupert protests. "He's just a baby . . ." "Best take a look at this," rumbles Growler stepping inside. Sir Jasper and Scrogg plainly weren't expecting this. They exchange troubled looks.

RUPERT TELLS WHY HE TOOK YUM

Yum hides behind the man called Wing.
"Ah!" Growlers sighs. "Poor little thing!"

The pals admit they set him free
To save him from his misery.

"You see!" Sir Jasper starts to scoff,
"They meant to cart the Yeti off!"

"A lift to China! Well Pong-Ping,"
Says Growler, "I must see this thing!"

When he sees Sir Jasper and Scrogg follow PC Growler indoors Yum squeaks and hops behind Wing. "A-ah! Poor little thing . . ." Growler starts. He corrects himself: "Harrumph! Now, Rupert, did you take this gent's wild . . . (He pauses and looks at little Yum) . . . animal?" "He's not wild!" Rupert repeats hotly. "He's a baby Yeti who ought never to have been captured. And, yes, I took him because he was so miserable." "So did I!" Pong-Ping cries.

"You see! Those two did take my Yeti!" snarls Sir Jasper. "Only to take him back to his home in the Far High Chinese Mountains!" Pong-Ping retorts. "Oh, and how were you going to get there?" Growler asks. "Why, by my private lift," Pong-Ping says. "A lift to China!" exclaims Growler. "Never heard of such a thing. Let's have a look at it!" He motions the Peke to lead the way. "That's very odd," thinks Rupert. "I *know* that Growler knows about Pong-Ping's lift."

21

RUPERT STARTS A LONG TRIP

The pals are stumped by Growler's drift
They're sure he knows about the lift.

The Peke displays his lift with pride
And shows they can all fit inside.

"Oh!" Growler cries, "For goodness' sake!
I've started it!" But by mistake?

As both the baddies gasp and blink
The lift then slowly starts to sink.

Pong-Ping leads the way to his lift, wondering like Rupert, why Growler is acting as if he has never heard of it. He enters the lift with Growler who makes a great show of gazing around. As he does so he rests his hand on a lever. "Hi, careful – that starts it!" cries the Peke. The bobby lifts his hand. "Surely you couldn't all get into this?" he says. "It's bigger than it looks," Pong-Ping replies. "Let's see" Growler says. The Peke beckons to Wing, Wang and Yum.

Nervously they step into it. It's a bit of a squeeze, but once they're settled there's still some space. "I *am* astonished," Growler admits. "Why, we could fit you in as well, Rupert," he marvels. "Let's try." Puzzled, Rupert squeezes in as Sir Jasper and Scrogg look on, wondering what Growler's up to. "Careful!" Growler gasps, "or you'll be pushing me against the starter – oh, oh, you have!" There is a humming and the lift begins to move!

RUPERT ARRIVES IN CHINA

"What have I done?" gasps Growler, yet,
Does not seem really so upset.

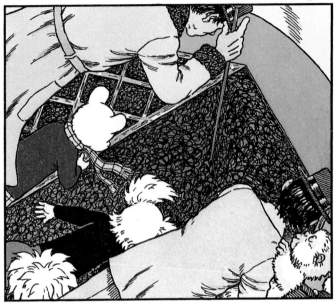

So that they come the right way out
Half-way the lift must turn about.

China! And towering to the skies
The mountain range where Yum's home lies.

Pong-Ping says Yum would like to know
If Growler plans to let him go . . .

"Oh, dear!" cries Growler as the lift speeds towards China. Yet he neither looks nor sounds as upset as you might expect. "What have I done?" he adds. "I know what you've done," thinks Rupert. "You've been a rather crafty and very kindly policeman." And he's sure Pong-Ping has the same idea. But now the Peke speaks in Chinese to Wing and Wang. "I've told them the lift turns over about now so that we reach China right way up," he explains. "Ready – now!"

At last the lift stops. "China!" Pong-Ping announces. Rupert and the others step out into daylight. "China!" PC Growler repeats. "Well, I never!" Just then Yum spies the distant mountains and squeaks with glee. Somewhere among them is his home. Then he stops, looks solemn and squeaks inquiringly. There is a burst of Chinese between Wing and Pong-Ping who turns to Growler. "They want to know if you are going to arrest them," he explains. Growler sighs.

RUPERT SHARES A JOKE

He says – and doesn't hide a grin –
In China he can't run them in.

"Nothing can stop you now, I'd say
From making good your getaway!"

"Escaping as I thought they would!"
Cries Growler, and then murmurs, "Good!"

"Now we had best get home, I think."
He tells the two pals with a wink.

"Oh, I can't arrest anyone here," Growler says in a regretful tone. "I'd have to get a Chinese bobby to do that and by the time I'd found one Yum and the others would be miles away." Rupert grins at Pong-Ping. "Tell them that," he says. "'Specially the bit about being 'miles away'." Pong-Ping laughs and addresses Wing and Wang. They smile broadly and bow to Growler and the pals before taking little Yum's hands and turning towards the mountains.

"Escaping, as I feared they would," sighs Growler as Yum, between Wing and Wang, dwindles into the distance. "We can tell everyone you did all you could," says Rupert with a straight face. "And you *did* follow the case right to China," adds the Peke. "So I did," agrees Growler. "Not bad for a village bobby really. Now I think we'd best all go home." So the three get back into Pong-Ping's lift. As they do, Growler winks at the others. The End.

RUPERT

and the
Sea Queen

RUPERT GOES ON HOLIDAY

The Bears have just arrived today
At Cocklemouth on holiday.

"Is that our cottage?" Rupert cries.
"Yes, just there!" Mr. Bear replies.

His mother says that Rupert may
Visit the harbour straightaway.

"A good idea!" says Mr. Bear.
"Let's go and see what boats are there."

M r. Bear is getting over a bout of 'flu and the doctor has said that a short break by the sea will help him get well sooner. So when he hears of a cottage to let in Cocklemouth, and since Rupert's school is on mid-term break, the Bears decide to have an unexpected holiday. Rupert has never been to Cocklemouth and he's agog as they emerge from the railway station. They find their cottage at the top of a street running down to the harbour.

To make the holiday even more fun Rupert's pal Bill Badger is joining the Bears next day. His regular visit to the dentist has prevented him travelling down with them. While Mr. Bear changes into holiday clothes Mrs. Bear unpacks Rupert's things in the room Bill and he will share. "Why don't you two have a look at the harbour while I get supper?" she suggests. Rupert thinks this is a splendid idea and a minute later he and Mr. Bear are making their way there.

RUPERT FINDS SOMETHING FUNNY

*Down at the harbour our two see
A sailing ship moored at a quay.*

*"How funny!" Rupert laughs with glee.
And calls, "Oh, Daddy, come and see!"*

*His father chortles, "Goodness me!
She looks as sour-faced as can be!"*

*They leave but Rupert says that when
Bill joins them he'll come back again.*

Rupert loves ships. He can make up stories about where they go and the adventures they have. So Cocklemouth harbour is just the place for him. In a quiet corner he spies an old sailing ship. "Oh, let's have a close look," he says and runs on ahead of Mr. Bear. When Mr. Bear reaches the ship he finds Rupert staring at something at the front. "I see she's named the Sea Queen, Rupert," he calls. "I know," laughs Rupert. "Come and look at her!"

"What's so funny about her?" asks Mr. Bear. "See for yourself!" insists Rupert. And when Mr. Bear does look he bursts out laughing. The figurehead of the ship is the head of a woman wearing a crown. And what a face! It's the sourest face the pair have ever seen. "She looks as if the water smelt awful," chuckles Mr. Bear. "Or as if she's eaten something that's disagreed with her," suggests Rupert. "Oh, I must bring Bill to see her tomorrow."

RUPERT GETS A SHOCK

Next morning Rupert greets his chum
And cries, "Bill I'm so glad you've come!"

"Just dump your case and follow me,
There's something I want you to see!"

Runing down to the quay once more
Rupert shows Bill the ship he saw.

"Wait 'til you see the figurehead!"
He starts to say then gasps instead . . .

Next day Rupert is at the station to meet Bill Badger's train. "Thanks for asking me," Bill greets him. "Sorry about not being able to come yesterday but I'd been booked for ages to see the dentist." "Well, you're here now and that's the main thing," laughs Rupert. "Now come and I'll show you the cottage. You can dump your case there and then I want you to see something." But he won't let on to Bill what it is, not even when they're hurrying to the harbour.

As they go Rupert has a sudden thought: "Oh, wouldn't it be awful if that ship has sailed." But when the pals reach the quay the Sea Queen is still in its quiet corner. "That's what we've come to see," says Rupert pointing to the ship. "The Sea Queen," Bill reads the name on the stern. "What's so special about it?" "Wait 'til you see the Queen," Rupert chuckles. But when they see the figurehead Rupert gasps and Bill asks again, "Well, what's so special?"

RUPERT IS ACCUSED OF LYING

The Sea Queen looks so different now!
She's pretty and she smiles. But how?

As Bill and Rupert stand and stare
An angry seaman hails the pair.

"The Sea Queen!" Rupert starts to say.
"Her face has changed since yesterday!"

The sailor scoffs 'til Rupert shows
He's not the only one who knows.

"But . . ." is all Rupert can say. The figurehead is smiling! "But it wasn't like that yesterday," he manages to say at last. And then he goes on to describe the sour-faced figurehead. "Daddy saw it too," he adds. The pals stare at the smiling wooden face, Bill confused and Rupert finding it hard to believe his eyes. Then an angry voice cuts in on their thoughts: "What are you two staring at?" The pals jump. A rough-looking man has appeared on the Sea Queen's deck.

"Oh, please, we didn't mean to be rude," Rupert apologises. "But we were jolly surprised." And he tells the man why they were. For a moment the man seems startled then he starts to scoff: "Figurehead started to smile, eh? Maybe you can tell me how a bit o' carved wood could do that. No? I didn't think so. Now be off! I can't waste time listening to silly lies." "They're not lies!" Rupert bursts out. "I saw it with my own eyes, the nasty face. So did my Daddy!"

RUPERT IS CHASED AND CAUGHT

He starts to smile and says the two
Should come and have a closer view.

But they don't trust the bearded man
And run off quickly as they can.

The two pals race along the quay.
"This way!" cries Rupert. "Follow me!"

To their dismay the pals then find
That someone grabs them from behind.

At the mention of Mr. Bear the man's manner changes. "Now, now! I'm sorry!" he cries bluffly. "I can see you're really not the sort to tell lies." As he talks he starts down the gangway. "I think you've just got mixed up," he goes on. "Now, why not come aboard and have a proper look at the old ship?" But Rupert and Bill are not taken in. As the man advances they retreat. Then as one they turn and run. With a roar of rage the man bounds after them.

As he hares along the quayside Rupert has the strong feeling that if the man catches them Bill and he are going to end up on the Sea Queen, like it or not. But the man is hampered by heavy seaboots and the pals are managing to stay pretty well ahead. "Let's try to lose him in here," Rupert pants and jinks into a gap between piles of bales and crates. It is beginning to look as if they're going to get away. Then they are grabbed by two large hands.

RUPERT MEETS THE CUSTOMS MAN

The Harbour Master's caught the pair.
And asks them what they're doing there.

Just then the sailor reappears.
"Those two are trespassers!" he sneers.

The Harbour Master's now more grim.
He says the pals must come with him.

But as the pair are lead away,
A Customs man asks, "Who are they?"

The owner of the large hands turns out to be a fat man in uniform. He does not look pleased. "I am the Harbour Master," he announces in a pompous voice. "And I'd like to know what you think you're up to, racing about in here." But before the breathless pals can answer, the man from the Sea Queen arrives. "I was chasing them," he pants. "I caught them on my ship, doubtless looking for anything worth stealing." The pals are quite speechless at the man's words.

"Trespassing on ships, eh?" the Harbour Master cries. "We can't have that. You'd better come along with me." And before they can protest the pals find themselves marched away while the man from the Sea Queen turns back to his ship. Near the Harbour Master's office a man in Customs Officer's uniform steps from a doorway. "Hello, Harbour Master," he calls. "Who have you there? Smugglers?" "It's not funny," the Harbour Master says. "They're trespassers."

RUPERT TELLS ABOUT THE FACE

*"We weren't trespassing! It's not true!
Just let me tell the facts to you."*

*He tells the officer the way
The Sea Queen's changed since yesterday.*

*The Customs man says. "If you're right,
They changed the Queen's head overnight!"*

*"I'm going to see it with this pair.
Get the police to meet us there!"*

"They don't look like villains," smiles the Customs Officer. "We're not!" cries Rupert who has got his breath back. Then before the Harbour Master can interrupt, he pours out the true story. At the mention of the Sea Queen the Customs Officer's smile fades. But when he hears why the pals were interested in the Sea Queen and about the change in the figurehead he becomes very serious. He drops on one knee beside Rupert. "Tell me again," he asks.

The Customs Officer listens intently to the account of how Rupert and Mr. Bear laughed at the sour-faced figurehead and how, when Rupert and Bill went to look, it was smiling. "A wooden figurehead can't change its face," the Customs man says slowly. "But it can *be* changed!" He turns to the Harbour Master: "Bring the police to the Sea Queen at once. I'll explain later. I'm going there now." To the pals he says, "You two come as witnesses."

RUPERT HELPS THE CUSTOMS MAN

*"We'll go at once, for I believe
The ship could be about to leave."*

*What has the Customs man in mind?
What is it he expects to find?*

*They drift alongside and the man
Says, "Hold us steady as you can."*

*But as he reaches for the Queen
It's evident their boat's been seen.*

The startled Harbour Master goes bustling off to fetch the police and the Customs Officer leads Rupert and Bill to where his nippy-looking launch is moored. "I must act at once," he says. "The Sea Queen could sail any time. I can't risk it going before the police get here. So I'll need witnesses to anything I find before they arrive. That's why I've asked you along." The pals follow him into the launch and in a moment are racing towards the Sea Queen.

Just before the launch reaches the Sea Queen the Customs man cuts the engine and swings his craft neatly under the figurehead. He passes a boathook to Rupert. "Catch on to that ladder behind you and hold the launch steady for me," he whispers. 'This is exciting," thinks Rupert. "What's he going to do?" What he does, in fact, is reach up, take the wooden head in his hands and start to twist it. At that moment a man's head appears over the ship's side.

RUPERT SEES AN ARREST

"Stop!" shouts the sailor, but instead
The Customs man unscrews the head.

At this the sailor turns and flees,
Then stops and gasps at what he sees.

Down on the quay the Sea Queen's crew
Are rounded up – the big man too!

The Customs man then settles down
To scrape the Sea Queen's painted crown.

Glaring down at the launch is the very man who chased Rupert and Bill. "Hi, stop that! What d'you think you're up to?" he yells. "Taking a closer look at this lady," the Customs man says. With that, he gives the crowned head a sharp twist then to the pals complete astonishment turns it a couple of times and lifts it off! As he starts up the ladder with it there is a cry of "Police!" from the ship. The Sea Queen man swings round and gives a gasp of dismay.

Now that there's no need to hold the launch steady Rupert and Bill follow the Customs man up the ladder. And what a sight greets them on the quayside! Two of the Sea Queen's crew, looking dismal, are standing between the Harbour Master and a policeman while the man who chased the pals is marched ashore by another policeman. "Take them and lock them up!" the Customs man orders. Then as they're marched off he sits down with the crowned head and produces his knife.

RUPERT LEARNS THE TRUTH

*The paint's soon scraped away to show
Bright gleaming gold and jewels below.*

*"The robbers took the old head down
And swapped it for a stolen crown!"*

*A moment later the pals see
Familiar figures on the quay.*

*The Customs man shows them the Queen
And tells them where the pals have been.*

The Customs Officer scrapes the paint on the crown. The pals' eyes pop as what they thought was wood is revealed as gold set with gems. "The crown of Ruvalia!" says the Customs man. "Stolen from the Ruvalian queen on a visit here. For state reasons the theft was kept secret. Customs Officers were told to keep a sharp lookout for an attempt to get it out of the country. When Rupert told me about the changed figurehead I saw how it was to be done. They were going to sail it out – looking just like a crown." "I'd not have noticed if they'd bothered to copy the old Sea Queen's face for the one they put in its place," says Rupert. "But they didn't and *you* did," laughs the Customs man. Just then Mr. and Mrs. Bear appear and are introduced to the Customs man. He sees them goggle at the head he is holding. "There's an exciting tale behind this," he tells them. "And your son Rupert and his pal played leading parts." The End.

RUPERT

*Awakened by a wooshing sound,
Rupert starts up and looks around.*

Rupert is in that not-quite-asleep state you reach sometimes if you've been dreaming when – woosh! – a sound like rushing wind has him wide awake. By the time he scrambles to the window and opens the curtains the sound has faded but he's in time to see something that glows in the sky before vanishing behind a belt of trees. He waits but nothing more happens and he climbs back into bed very puzzled indeed.

and the Eastern Isle

Quick! To the window! There he sees
A strange light drop behind the trees.

Next day at school his pals all tell
How they saw that strange light as well.

At school next day Rupert finds he isn't the only one to have seen the strange light. Everyone is talking about it. Falling star? Space rocket? The arguing and guessing stop only when Dr. Chimp rings the bell for class. But before the first lesson he opens a note he has received. "Tigerlily will be off for a while," he says. "Her father, the Chinese Conjurer has to take her away unexpectedly." Just then Pong-Ping rushes in.

Their teacher gets a note to say
That Tigerlily's gone away.

RUPERT HEARS OF A LOST PET

*Pong-Ping was late because he feared
His dragon pet had disappeared.*

*"His food's still here!" exclaims Pong-Ping.
"He hasn't eaten anything!"*

*Then as the two pals hunt around
Rupert calls, "Hey! Look what I've found!"*

*The garden hedge has been burnt through.
The grass scorched on the far side too.*

As they walk home later Pong-Ping tells Rupert why he was late for school. "I was looking for my little pet dragon. It was missing when I got up today. I searched for as long as I could, but it got so late I left out a bowl of its favourite food and ran all the way to school. I'm worried." "It will probably be there when you get home," Rupert says. "Look, I'll come with you." But when they get to the Peke's house the pet dragon is still not there and its food is untouched.

Now Pong-Ping really is worried. "Then we must search for it at once," Rupert decides. So they split the big garden between them and begin. It is Rupert who spots what could be a clue. "Hey! Look what I've found!" he calls and points to a scorched hole in a hedge. "That's it!" the Peke cries. "Whenever my dragon gets excited it forgets its training and breathes fire." The pals squeeze through the hedge and there on the other side is a trail of burnt patches.

RUPERT FINDS THE PET

The scorch marks lead the pals to where
Great puffs of smoke rise in the air.

Pong-Ping's sure it must be his pet,
But finds the cause is stranger yet . . .

Another dragon's standing there
And blowing smoke clouds in the air.

It looks quite gentle. Even so,
Pong-Ping thinks they had better go.

Following the trail, Rupert and Pong-Ping set off after the pet dragon. The trail seems to lead straight to the tall tower where their friend Tigerlily lives with her father the Chinese Conjurer. Then from behind trees near the house they see puffs of smoke rise. "My dragon's there all right!" cries Pong-Ping. But to Rupert it seems a lot of smoke for one small dragon to make. The pals push their way through some bushes. The pet is there. But not alone.

The Peke's pet is smiling fondly up at a huge version of itself. Rupert and Pong-Ping goggle at it. It looks back, but in the gentlest sort of way. Very cautiously the Peke advances and lifts his pet. "Let's get out of here and tell someone about this," he whispers to Rupert. But as the pair of them steal away, the big dragon looks terribly sad as if it were losing a very dear friend and so does the little dragon which is making little whimpering noises.

39

RUPERT IS CAPTURED

"That's Tigerlily there! I say!
But she's supposed to be away!"

She waves, then gives a sudden shout
Of warning to the pals, "Look out!"

Too late the chums swing round and see
An armoured guard loom threateningly!

He bundles them inside to where
Poor Tigerlily's in despair.

"I suppose we'd best go to the village and tell PC Growler," says Rupert as the pair make their way back past the Conjurer's tower. "We could have told Tigerlily and her father if they'd been at home . . ." He breaks off with a gasp of surprise, for they are close enough to the tower to see through the windows and in one of them stands Tigerlily! He cries out in surprise and points. The pals call to her and Tigerlily throws open the window. But as she does, her smile of greeting vanishes and she screams, "Look out!" Rupert and Pong-Ping swing round to find a man in strange armour looming over them. Before they can do a thing he has grabbed them. They yell and squirm and Pong-Ping's dragon snorts smoke at the man. But all he does is bundle them effortlessly into the Conjurer's house and thrust them into the room where Tigerlily is standing. Then without a word, he leaves. "Now you are captives too!" Tigerlily wails.

RUPERT MUST GO A LONG WAY

*"Last night two men like that came here
And locked us up. Though why's not clear!"*

*But as she speaks the guard says they
Must follow him without delay.*

*"They're taking me away. And you,"
The Conjurer says, "must come too!"*

*How will they go? Our lot soon see.
By dragon, it would seem to be.*

"What is all this?" gasps Rupert. "Who's that man? Why are you here? Teacher got a note from your father to say you were going away for a while." "That man and his friend must have made my father send it!" cries Tigerlily. "They came in the night and made us prisoners in our own home. I don't know why." Just then the man returns. "All come!" he orders. And with that he bundles Rupert and Pong-Ping out of the room with Tigerlily following.

Outside, Tigerlily's father is waiting with the second of their captors. "I am sorry," he tells the pals. "These men say I must go to their country with them. Now they say that you must come too for no one must know where we have gone." "Oh, dear! Is it far?" quavers Rupert. The Conjurer nods. "How do we go?" Pong-Ping asks. One of the men seems to understand for he speaks to the other one who hurries away. When he returns he is leading the big dragon.

RUPERT TRAVELS BY DRAGON

The dragon's made to bow down low.
A guard says curtly, "We go so!"

On mighty wings and breathing fire
The dragon struggles ever higher.

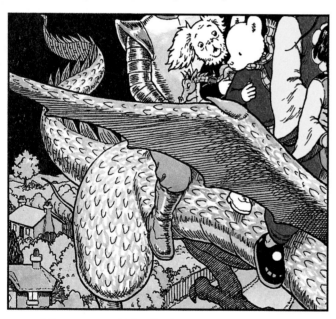

That strange light flashing through the skies
Last night was this, they realise.

Soon Nutwood passes out of sight.
They fly on Eastward through the night.

At a command from one of the men the big dragon crouches and the other climbs on to its back. "We go so!" says the first man and beckons to the Conjurer who helps Tigerlily up and takes his place behind her. "Now you," the man growls at Rupert and the Peke. Rupert doesn't fancy arguing and does as he's told. But Pong-Ping has to be bundled aboard followed by his pet. Then the man follows them, an order is given and with a "woosh!" the dragon takes off.

Breathing fire with the effort of rising so heavily laden, the big dragon soars over Nutwood. Below, the lights are coming on. "I see now!" Pong-Ping shouts above the rush of air. "See what?" Rupert asks. "What that strange light in the sky was last night," the Peke replies. "It was this thing arriving." Familiar countryside is soon left behind and the four Nutwood captives find themselves over wide seas and towering mountains, heading eastwards.

RUPERT REACHES A PALACE

At last the sun comes up to show
A range of mountains there below.

Beyond the mountains next they see
A palace – splendid as can be!

The dragon lands and straight away
An armoured soldier runs their way.

The friends dismount and find that they
Are marched inside without delay.

On through the night the dragon carries its load of unwilling passengers. The guards speak only when spoken to and even then answer curtly. All Rupert is told when he ventures to ask how long the journey will take is, "Until the sun rises." At last the sky ahead begins to lighten and the sun appears over a range of mountains. Beyond the mountains stretches a sea. On a headland stands a magnificent palace. The dragon glides gently down towards it.

Without a word from the guards the dragon heads for a courtyard in the palace and touches down there. The guards dismount and beckon to the others to follow. As they climb down, a soldier carrying a battleaxe runs up. The guards speak to him in their own language then lead the big dragon away. It casts a sad look back at Pong-Ping's pet. The soldier marshals the prisoners into a line led by the Conjurer and marches them up a flight of stairs and into the palace.

RUPERT MEETS THE RULER CHAN

In one huge chamber, quite alone,
The ruler Chan sits on his throne.

The Conjurer must find, says Chan,
The secret of the isle, Tang-San.

When Chan's told no such isle can be,
He leaps up hissing, "Come with me!"

He leads the little party to
A terrace with an ocean view.

As they are marched through the palace Rupert tries to imagine what lies ahead. He finds out in a huge chamber where their escort halts them at the foot of a high throne on which sits the fattest, most unpleasant looking man Rupert has ever seen. "Bow!" commands the soldier. "Bow to Chan the Magnificent, Ruler of All the East!" "Not quite all," hisses the fat man. "Magical Tang-San is not mine. And that is why you are here, Conjurer – to get me its secret!"

"Tang-San." The name means nothing to Rupert. Plainly, though, it does to the Conjurer. His tone is respectful but he smiles slightly as he addresses Chan: "The magical isle of Tang-San is a mere legend, a myth. There is no Tang-San." Rupert would never have imagined that anyone so fat could rise as quickly as Chan at the Conjurer's words. "No? Then come," he hisses. And he waddles out of the chamber at the head of the four friends, heading for a terrace.

RUPERT SEES THE MAGIC ISLE

"Tang-San!" he cries, but Rupert's sure
The island wasn't there before!

That isle, says Chan, the friends should know,
Can, as it chooses, come and go.

To Tang-San Chan's sent his best men,
He says, but none's come back again.

"Your daughter is my guarantee
You'll bring its secret back to me!"

On the terrace which overlooks the sea Chan points to a small island. "I didn't see *that* when we flew in," murmurs Rupert. Chan hears him. "Did *you* see it, Conjurer?" he demands. "No," admits the Conjurer. "Because it was not there then," hisses Chan. "It comes and goes as it chooses. Between when you landed and were brought to me I received word from my lookouts that the magic isle had come back. Behold the isle you say does not exist!"

"Haven't you sent any of your men to find out about the isle?" At Rupert's words Chan turns. "When Tang-San has stayed put long enough I have sent some of my bravest officers. None has come back." He turns back to the Conjurer: "So I have decided to match magic against magic. You shall bring Tang-San's secret to me!" "You cannot make me," the Conjurer declares. "I can," smirks Chan. He points to Tigerlily. "My guarantee."

RUPERT MUST STAY BEHIND

He's furious, but even so
The Conjurer sees he must go.

And as he leaves he stops to tell
The little ones, "All will be well!"

There seems no way to help him, yet
Pong-Ping plans something with his pet.

"May my pet dragon serve you, please,
And go with him?" Yes, Chan agrees.

"Your daughter and her friends shall be my prisoners until you return from Tang-San with its secret." Chan sneers. The Conjurer seldom shows his feelings but Rupert has never seen anything as chilling as the look he turns on Chan. When he does speak he says only, "I shall go." Chan beckons a guard who hurries across to loom over Rupert and his pals. All this is very frightening. "Be still, little ones," the Conjurer says quietly. "All will be well."

Rupert wishes he could believe "all will be well". He looks at Pong-Ping to see how he is taking it. But the Peke's attention is on his dragon. He seems to be making up his mind. Then he chirps at his pet in what Rupert knows is dragon language. The pet chirps in reply. Now Pong-Ping addresses Chan: "O, Great One, may my dragon serve you by accompanying the Conjurer as helper and messenger?" The fat ruler stares at the pet. "Oh, very well,' he grunts.

46

RUPERT'S PAL HAS A PLAN

The pals are marched off. As they go
They see a waiting boat below.

Then high up from a bridge our three
Can see the boat put out to sea.

They wave the Conjurer farewell
Then they are marched down to a cell.

Pong-Ping explains how he will use
His pet to bring them any news.

As soon as Chan has turned back to his throne room Rupert, Pong-Ping and Tigerlily are marched off by one guard and the Conjurer and pet dragon by another. Before they part the Conjurer says, "We shall be together again soon." As they are led from the terrace Rupert can see the Conjurer being taken to a boat. A few moments later as he and the others are ushered across a bridge they see the boat heading for Tang-San – which is staying put.

Rupert and the others are marched upstairs and downstairs and along passages until they reach a forbidding looking door which the guard unlocks. He ushers them into a cell. It is stark and comfortless but it does have a window – with bars, of course. As soon as the guard has locked them in Pong-Ping explains what passed between him and his pet in dragon talk: "As I told Chan, it can act as the Conjurer's messenger. But not to Chan, as he believes. To us!"

RUPERT'S PAL'S PLAN WORKS

Then at the window there's a face –
The dragon they rode to this place!

It peers inside but doesn't stay:
It sees the pet has gone away.

Later the pals are wakened by
The little dragon's chirping cry.

Outside the window of the cell
They see the big one's there as well.

Pong-Ping has just finished explaining that his pet will be able to act as a messenger if the Conjurer finds something he wants only them to know or if he thinks of a way of freeing them, when the light from the window is cut off. The pals swing round to see the big dragon that brought them from Nutwood peering anxiously into the cell. It gulps dismally then flies off. "It was looking for my pet," says Pong-Ping. "They're terribly fond of each other."

At last night falls and the three pals, having finished the supper brought by a guard, settle down to sleep. They are dozing off when they are startled awake by chirping. There at the window is Pong-Ping's pet. Next moment it and the Peke are chirping excitedly to each other. "It has come to take us to your father, Tigerlily," explains Pong-Ping. "But how?" Rupert asks. "Look outside," says the Peke. The pals crowd to the window. The big dragon is there.

RUPERT ESCAPES BY DRAGON

It bends the window bars agape
So that the pals can all escape.

Pong-Ping explains the dragon's plan.
It's going to take them to Tang-San.

The dragon flaps its wings and they
Soon clear the walls and fly away.

"My pet," states Pong-Ping, "Won't tell me
What Tang-San's like – says, 'Wait and see!'"

"I don't know how they were able to talk to each other," says Pong-Ping, "but my pet convinced the Conjurer that the big dragon is so fond of it that it might be willing to help us escape. Plainly it is. It will fly us to Tang-San where the Conjurer awaits us." He chirps to the big dragon which grasps one of the window bars, pulls it from its socket and bends it aside. The other bar follows and next moment the pals have scrambled aboard their large rescuer.

Effortlessly the big dragon takes off with the three pals and the little one on its back. As they clear the palace roofs they look down on the startled faces of Chan's sentries. In a moment they are over the sea. "Pong-Ping," says Rupert, "ask your pet what Tang-San is like." There is an exchange of dragon chirps. The Peke looks surprised when he turns back to Rupert. "For some reason it will say only that we must see for ourselves," he reports.

RUPERT SEES THE ISLE MOVE

Rupert suspects that something's wrong.
This journey's taking far too long.

Then Rupert gives a cry, "I say!
I'm sure Tang-San has moved away!"

The pet forgot it had to show
A signal to the isle below.

So now it breathes out flames until
The magic island stops quite still.

"Oh, well," says Rupert, "we'll be able to see for ourselves soon enough. We'll be there soon." But even as he speaks it strikes Rupert that they're taking an awfully long time to get to Tang-San. It seems to get no nearer. In fact, it seems almost further off than before. He looks back to see how far they have come and with a shock sees the land they have left dwindling in the distance. Then the truth dawns on him. "I'm sure Tang-San has moved away!" he cries.

At Rupert's cry of dismay the pet dragon starts to chirp excitedly. Pong-Ping translates: "It says that in all the excitement it forgot it has to make a signal so that the island will know it is us approaching. For all the island knows, Chan could be on this dragon. After all, it is one of his." By now the little dragon is perched on the other's head, signalling with bursts of flame. With a shout of glee Rupert and the others see Tang-San come to a stop.

RUPERT REACHES THE ISLE

The pals are baffled by the way
The island saw their sign to stay.

The island's rocky ramparts hide
A lovely landscape tucked inside.

Next moment Rupert finds he may
Understand all the dragons say!

"Land on the lake!" the small one cries.
"Impossible!" its friend replies.

As they draw nearer Tang-San Rupert gets the odd feeling that not only has the island stopped but that it seems to be welcoming them. "Pong-Ping," he asks, "did your dragon really say that the *island* didn't know who we were?" The Peke hesitates, then says, "Yes, it did, didn't it." But there is no time to ask the little dragon what it meant for they are crossing the steep rocky ramparts that ring Tang-San. Below, a lake gleams in the moonlight.

Rupert, Pong-Ping and Tigerlily are too awed to speak as they see the strange, beautiful landscape that lies behind the rocky ramparts. But the little dragon is chirping to its large friend. Rupert supposes it is giving directions. Then a most wonderful thing happens. As they dip below the ramparts Rupert finds that he is able to understand what the dragons are saying! "Land on the lake," directs the little one. "What? The lake? Impossible!" the other cries.

RUPERT LANDS ON SOLID WATER

"Go on!" it chirps, "This is Tang-San.
Land on the lake. I know you can!"

They land and as they do they see
The lake's as solid as can be!

Although the lake supports them, they
Just can't see how it changed this way.

"It changed," the pet says, "for you see
Here things are as you'd have them be!"

"What's happening?" gasps Tigerlily. "Suddenly I can understand dragon talk." "So can I!" Rupert cries. "And I don't like what I hear!" "Yes, what are you playing at?" Pong-Ping demands of his pet. "Land on the lake! Dragon's can't land on water!" But the pet says only, "This is Tang-San." And it repeats its direction to the big dragon. The big one gulps, glides slowly down to the lake. Hovers. Lands. No splash. The lake is like solid glass!

There is a long amazed silence. Tigerlily is the first to speak, voicing what the others are thinking: "But I saw the lake's surface move as we came down." Then nervously they dismount onto the glassy surface. Not only does it support them. They can even jump on it. "Yet it looked so like water," Rupert breathes. "It *was* water then," says the small dragon. "But for the moment we don't want it to be. And this *is* Tang-San, the marvellous isle."

RUPERT MEETS CHAN'S SOLDIERS

Then as the three pals reach the shore
The lake is as it was before.

They turn to go but find a pair
Of Chan's soldiers are standing there!

But what is this? Chan's soldiers smile
And welcome our three to the isle.

Then, as the pals give greeting too,
A voice says, "Tang-San welcomes you!"

Daylight breaks as the little party cross the solid lake to the shore. As they are about to leave it Rupert tosses a pebble at the lake. It sinks with a splash! The pet dragon smiles at his amazement and says, "If what you want is good or worthwhile then things on Tang-San are as you want them. We wanted a solid place to land. But no longer." Just then the big dragon cries out in alarm. Two men have appeared. "Chan's fiercest officers," the dragon quavers.

Rupert and the others stare miserably at the two soldiers bearing down on them. Rupert remembers Chan's saying he'd sent some of his best men to Tang-San. But wait! The men are smiling. "Greetings, little ones," they beam. "Greeting to you!" the chums chorus. And the odd thing is they find they mean it. Tang-San does seem to have the strangest effect on things and people. Then from somewhere close by a gentle voice says, "Tang-San welcomes you."

RUPERT LEARNS ABOUT TANG-SAN

The three pals turn around and stare.
The Chinese Conjurer is there!

He didn't speak. It was this man
Who greeted them – Voice of Tang-San.

"Tang-San itself speaks, truly so,"
Voice says. He's only there for show.

The name Tang-San means, "Isle of Good",
Here everything thrives as it should.

The chums turn to see who spoke. The Conjurer is there. "Daddy!" Tigerlily cries. "Was it you who spoke. It did not sound like you." Her father smiles, shakes his head and indicates a young man sitting nearby. "It was I – Voice of Tang-San," the young man says. "Please, are you the owner?" ventures Rupert. "No one owns Tang-San," replies the other. "I am only Voice – something to see for those who find it hard to believe the island itself truly speaks."

Rupert can't begin to understand how an island can move and speak. Yet he is sure he is being told the truth. The young man speaks again: "In our very ancient tongue Tang-San means the Isle of Good. As you have seen it may move where it chooses. The climate is always pleasant. There is beautiful fruit to eat and pure sweet water to drink. See." And he rises to show the chums a pool of sparkling water with lush, glowing fruit hanging over it.

People who come here find that they
Can understand what creatures say.

It's perfect here – he can't believe
That anyone would want to leave!

But Rupert says he'd rather be
In Nutwood with his family!

He'd be unhappy in this land,
Tang-San, Voice says, does understand.

"Here," Voice of Tang-San goes on, "each person and thing understands the other." He turns to the big dragon: "Is that not so?" "Oh, yes!" the dragon says. "And here," Voice says, pointing to the soldiers, "all is peace. They were Chan's fiercest warriors but deep down they wanted to be kind. Tang-San made them see that. They would never leave now. But who *would* want to leave so perfect a place?" "Oh, please," says Rupert in a small voice, "I do."

"You *want* to leave here?" Although Voice of Tang-San looks as calm as ever, it's plain he's surprised to say the least. "Please, I don't mean to be rude," Rupert says. "I'm sure Tang-San is all you say. But Nutwood *is* my home. Mummy and Daddy are there. It isn't perfect like Tang-San. Everyone isn't good all the time and often the weather is awful. But it's where I want to be." A gentle smile breaks over Voice's face. "Tang-San does understand," he says.

RUPERT RETURNS TO NUTWOOD

Since Rupert loves his own home so
It's back to Nutwood now he'll go.

Next minute Rupert stands and blinks
On Nutwood Common. "Home!" he thinks.

He'll miss his chums like anything.
He sighs. But look! There stands Pong-Ping!

He says, with Rupert gone, he saw
He'd have to come back here once more.

"You must love Nutwood," goes on Voice of Tang-San. "That is good. And here, as you know, if what you wish is good . . ." He breaks off then adds, "Close your eyes." Rupert obeys and misses the looks on his friends' faces, looks of dismay. Then he hears Voice say, "Open them." He does and cries out in amazement. He is on Nutwood Common! Home! But in a moment his delight fades. "My friends," he sighs. "Of course they will want to stay on Tang-San."

Happy though he is to be home, Rupert feels pretty dismal as he makes his way down from the common. So many friends left behind on Tang-San. With a sigh he sees Pong-Ping's house come into sight. Then he stops and goggles. For who's standing there grinning but the Peke with his pet dragon. Seconds later Pong-Ping is saying shyly, "Well, home's *here*, isn't it? And Voice said he could see I'd much sooner be here with those I'm fond of. So . . ."

56

RUPERT'S FRIENDS JOIN HIM

*"My pet agreed he'd also come
If he could visit his new chum!"*

*And are the others home once more?
No! No one answers their front door . . .*

*Their names are called. The pals swing round.
There stand the others, safe and sound.*

*"We left Tang-San and came back too.
We couldn't stay there without you!"*

Since two of them have decided to return to Nutwood maybe Tigerlily and the Conjurer have as well, Rupert and Pong-Ping agree. So they make their way to the Conjurer's tower. On the way Pong-Ping explains that, pals though his pet and the big dragon are, everyone thought it best for the big one to stay on Tang-San and for the two to meet from time to time in Happy Dragon Land. Tang-San would arrange it. But at the Conjurer's home disappointment awaits. The pals knock and knock but the door stays shut and they turn glumly away. Then – "Rupert! Pong-Ping!" The pals swing round. And there in the doorway are Tigerlily and the Conjurer. "We got back this very second," Tigerlily calls. "We couldn't stay behind without you." Her father smiles and adds, "Voice of Tang-San says Nutwood must be wonderful since we all wanted to return to it. The others burst out laughing and chorus, "Nutwood *is* wonderful!" The End.

These two pictures of Rupert arriving at the Chinese Conjurer's house look identical, but there are eleven differences between them. Can you spot them all? (*Answers on Page 99*)

RUPERT
and the
Little Train

RUPERT HAS A SPILL

The match is such a lively one
That all the pals are having fun.

Before the game is over though,
Rupert cries, "Sorry! Got to go!"

"I mustn't be late home today!"
He thinks and hurries on his way.

But as he takes a shortcut he
Falls over something he can't see.

It's a good game. Bill Badger and Algy Pug versus Rupert and Willie Mouse. Neither side is giving anything away. Then Algy bungles a pass, leaving Rupert and Bill to race for the ball. Rupert is almost on it when he stops stock-still with a look of dismay on his face. He seems not to hear Bill's gleeful shout or Willie's cry of "Wake up!" "Oh, gosh!" he groans. "I forgot!" Then with a cry of "Sorry! Got to go!" he sprints off towards Nutwood village.

The others are mystified by Rupert's sudden departure. What they don't know is that he has just remembered that his Mummy told him to be home especially early for lunch as she has a lot to do this afternoon. Now he's sure he is badly late; so sure that when he sees a likely shortcut through a wood he takes it. The grass in here is long and Rupert hasn't gone far when he trips over something hidden by it and goes sprawling headlong.

60

RUPERT FINDS THE TRAIN

"That's all I need!" poor Rupert groans –
At least he's got no broken bones!

That something's hidden there is plain.
He parts the grass and finds – a train!

It's not a toy, that much is clear.
But what on earth's it doing here?

He can't just leave it there and so
Decides to Growler it must go.

"Oh, that's all I need when I'm so late," groans Rupert. Luckily, though, apart from a bang on the ankle he isn't hurt and he picks himself up, meaning to hurry on. But, late though he thinks he is, his curiosity about what he tripped over is too much for him to resist and he turns back to the thick clump of grass where he stumbled. He parts it and gives a whistle of surprise at what meets his eyes. Standing there is a little train.

Rupert's rush to get home is forgotten – for the moment anyway – as he gazes at the train. It plainly is not a toy. It is perfect in every detail. "But who owns it?" wonders Rupert. "And what's it doing here?" Then there's the question of what he's to do about it. Anyone can see it must be worth a lot and shouldn't be left out here. But it's too bulky to carry. So Rupert decides that after lunch he will fetch a barrow and wheel the train to Nutwood police station.

RUPERT'S DADDY SEES THE TRAIN

Then Rupert hurries home to tell
His parents of the train as well.

"I must see this!" says Mr. Bear
"Come on, we'll take a barrow there."

"It is *a beauty! I agree,"*
Says Mr. Bear, "Real coal, I see!"

The carriage doors all open wide
And show it's got worn seats inside.

As it turns out Rupert isn't as late as he feared. Lunch has only started when he bursts into the house. "Just in time," smiles Mrs. Bear. "Sorry," Rupert puffs. "We were playing football and I forgot about having to be home early and I ran all the way and I fell over something and just wait 'til you hear what it was!" Well, when Rupert has told his story Mr. Bear announces, "I must see this." So after lunch he and Rupert set out with a wheelbarrow.

"You were right," Mr. Bear tells Rupert when he sees the little train. "It *is* a beauty!" He kneels to examine it. "Amazing!" he breathes as, from the bunker at the back of the engine, he takes a handful of what look like small black stones. "Real coal," he says. "And not just for show – there's ash in the fire-box. This is a *real* engine!" He's even more impressed when he looks inside a carriage. "Those seats!" he says. "I could swear they've been sat on!"

62

RUPERT SEES GROWLER UPSET

They cover up the barrow so
The train's protected as they go.

Then off they trundle straightaway
To see what Growler has to say.

"A train?" cries Growler. "Let me see.
You're right! It looks quite real to me."

Of course he'll take the train they've got.
On second thoughts, he'd rather not.

"Do you really think someone has been *riding* in this train?" asks Rupert. Mr. Bear hesitates. "Well –," he begins. He pauses then goes on briskly, "No, it's silly even to think such a thing. Come on, let's get it to the police station." So they load the little train into the wheelbarrow, cover it with an old blanket and start back to Nutwood. "You'd never have pushed this on your own," puffs Mr. Bear as at last the police station comes in sight.

"I bet PC Growler's never had to look after a lost railway train before," says Rupert as he goes in to get him while Mr. Bear waits outside with the barrow. In a moment Growler emerges exclaiming, "A train, you say. Let me see!" Mr. Bear pulls aside the blanket. "I see what you mean about a *real* train," muses the bobby. "We'll turn it over to you, then," smiles Mr. Bear. "Of course –," Growler starts. He stops. "I – I'd rather not," he stammers.

RUPERT TAKES THE TRAIN HOME

His youngsters scamper out just then;
He whisks the blanket back again!

"They'd want to play with it, you see
Can't you look after it for me?"

Our two agree that Growler's right.
They'll keep the small train for the night.

The best place for the train's inside
The garden shed, they both decide.

Rupert and Mr. Bear are taken aback by Growler's strange attitude. But before they can speak they hear the two young Growlers coming downstairs. "It's them," Growler whispers. "They'd make life a misery, wanting to play with the train and me having to say, no. Can't *you* look after it for now?" He whisks the blanket into place just as the small Growlers appear. Mr. Bear nods and next thing Rupert and he are heading home with the little train.

"Poor PC Growler," sympathises Mr. Bear as he trundles the wheelbarrow to their cottage. "I can well imagine how awful it would be for him with his youngsters forever pestering him about the train." "And I can imagine how the young Growlers would feel," Rupert chuckles. "So we'd better not put the train in *my* toy cupboard. "It wouldn't fit, anyway," Mr. Bear says. And he decides in the end that the best place for it is his garden shed.

RUPERT INTERRUPTS A RAID

The garden shed is locked up tight.
The train is safe there for the night.

But that night Rupert's wakened by
A scuffling sound, a high-pitched cry.

He peers out at the shed and then
Cries out as he sees – four wee men!

The men, who all have climbing gear,
Hear Rupert's cry and freeze in fear.

Rupert helps Mr. Bear unload the little train and stow it in a warm, dry part of the garden shed. Then Mr. Bear padlocks the door securely. Rupert hurries indoors to tell Mrs. Bear what's been happening. She laughs when she hears why PC Growler has asked Mr. Bear to look after the train. "Well, I'm sure it will be quite safe here," she says. But – is she right? Late that night Rupert is wakened by a noise. It comes from the direction of the garden shed.

Rupert is sure the sound was a squeal. He holds his breath. He can hear scuffling. He steals to the window and opens it. For a moment he can't believe what he sees. On the moonlit ground is sprawled a tiny figure which seems to have tumbled from a cord fixed to the shed's window-sill. Two others are making as if to help it. A fourth, carrying a tiny climbing-axe is on the window-sill. At Rupert's cry of astonishment they turn stricken faces on him.

RUPERT MAKES A FIND

"*Whatever's up?*" *asks Mr. Bear.*
"*Quick!*" *Rupert cries,* "*just look out there!*"

Now, what, he thinks, can Rupert mean.
There's simply nothing to be seen.

Next day he finds his parents seem
To think the whole thing was a dream.

"*But look, the axe they used last night!*
Those little men were here, all right."

For a long moment Rupert and the tiny figures stare at each other. In that moment Mr. Bear bursts into the bedroom, summoned by Rupert's shout. "Quick, Daddy!" urges Rupert. "The shed! Look!" Mr. Bear goes to the window. "Well," he asks, "what about the shed?" Rupert stares. In that instant he turned from the window the intruders have gone. He pours out his story but Mr. Bear merely smiles and says, "A dream, old chap. That's all it was. Now, back to bed."

Next morning Rupert still insists that the tiny men he saw were no dream. And he's sure they were trying to break into the shed. But his parents won't be convinced he wasn't dreaming. Thinking how awful it is when people will not believe what you *know* is true, Rupert goes out to the garden shed. And there he finds the proof he needs. The cord the tiny men were climbing still hangs from the window-sill and below it lies the climbing-axe one of them carried.

RUPERT LOOKS FOR CLUES

"It was no dream! See what I've found!
Beside the shed, left on the ground."

His father takes the axe and stares:
"Perhaps the little train is theirs?"

The next thing Rupert wants to do
Is try to find another clue.

He goes back to the common where
The mark left by the train's still there.

"Mummy! Daddy! I was right! It wasn't a dream! Look!" Rupert rushes indoors flourishing the miniature climbing-axe. His parents stare, then Mr. Bear takes the little axe and studies it. He gives a low whistle. "Like the train, this is not a toy," he muses. "Sorry we didn't believe you, old chap. So there *were* little men trying to break into the shed where we put the little train. Might it be . . . is it just possible – that the train is really *theirs*?"

The next question is what should they do now. "Well, I know what I'm going to do," announces Rupert, making for the front door. "I'm going back to where I found the little train." His parents follow him to the garden gate. "I'm going to look for clues," he tells them. "We didn't really look around when we were there." So he makes his way to the wood where he fell over the train and there he finds the place he wants, the grass still flattened by the train.

67

RUPERT IS TAKEN PRISONER

*The grass there is so long and thick
He has to clear it with a stick.*

*He whacks a clump and right away
There comes a loud cry of dismay.*

*"An Imp of Spring!" he cries. "Oh, no!
I didn't mean to hit you so!"*

*When Rupert tells why he is here
The Imp cries, "Seize him!" More appear.*

Rupert examines the patch of flattened grass but it yields no clues to why the train was there. "Such an odd place for it to be," he muses. "No rails, nothing. Yet plainly it hadn't been there long. Let's have a look around." He picks up a slim whippy branch from the ground and starts to whack aside the surrounding long grass in search of clues. Suddenly he leaps back and drops the branch when one whack of it produces a howl of pain.

Before Rupert's horrified gaze out of the long grass springs an extremely angry little figure which he recognises at once as one of the Imps of Spring. "What d'you think you're at?" the Imp screams. "S-sorry!" Rupert stammers. "I didn't mean . . . the train . . . I was . . ." But the Imp interrupts. "Aha! The train! You're on *their* side, eh? Imps, seize him!" In an instant more Imps of Spring appear as if from nowhere and surround Rupert.

RUPERT HEARS ABOUT THE RAILIES

"I wonder what's upset them so?"
Thinks Rupert as he's marched below.

They stop at last. Then off one goes
To fetch their King who Rupert knows!

"By rescuing that train of theirs
You've helped those Railies!" he declares.

"The Railies' train runs on this track
And turns our walls and ceilings black."

Rupert more than once has met the Imps of Spring whose job is to get things growing again after each winter. They're easily upset. Rupert wonders what has so upset them now. He suspects that it was his mention of the train. They drag him to a trapdoor, until now disguised as grass, and force him underground where one runs ahead, taps on a door and talks rapidly in an undertone. The door is opened by someone Rupert knows – the King of the Imps.

"So you have helped the Railies by rescuing their train!" The King's opening words take Rupert aback. Nervously he stammers out how he found the train and where it is now. "But I've no idea what Railies are," he pleads. The King ponders. "I believe you," he decides. "Come!" He leads Rupert to a rail line snaking between holes in blackened walls. "Railies did that with the train you found," he says. Rupert suddenly has an idea who the Railies might be.

"We ambushed them the other day
And got their train. They ran away."

"The Railies tried to raid our shed
But we surprised them and they fled."

"You didn't help the Railies so,
My Imps and I will let you go!"

He does not see as home he hares
The sly smile that the Imp King wears.

"Railies!" the King storms on. "No respect at all for others. Put their railway anywhere. Knock holes in walls for tunnels. All so they can joyride that train. But we showed 'em. Ambushed it. The Railies escaped but we got the train. We dumped it outside meaning to lie in wait and see if they came for it. But you moved it before we could." "Please, are Railies like tiny humans?" Rupert asks. The King stares at him. "Now, how do you know that?" he demands.

So Rupert tells the King about the tiny people trying to get into his shed. "And you say the train is there?" the King asks. Rupert nods. The King says nothing but plainly he is thinking. At last he speaks: "Well, it's clear you have nothing to do with the Railies and that you took the train innocently. My Imps got quite the wrong idea when they found you where we left it. So we shall let you go." But the King's smile as he sees Rupert off a little later is sly.

RUPERT IS STARTLED AWAKE

Rupert runs home without delay
And meets his father on the way.

"You say those Railies own the train?
We must see Growler and explain."

"I'll keep watch for those little men
Tonight in case they try again."

Much later Rupert wakes to hear
The sound of scuffling somewhere near.

Rupert rushes home to tell about his adventure with the Imps of Spring and arrives at the same moment as Mr. Bear. He's been into the village to tell PC Growler about the little people who tried to get into his shed. When he has heard Rupert's tale he says, "Well, no matter how they upset the Imps, the Railies *are* the train's owners. We must let PC Growler know. But that will have to wait 'til tomorrow. He was just off to Nutchester when I saw him."

At bedtime Rupert says to Mr. Bear, "If the Railies do own the train why don't they just ask us for it? They seem to know where it is." "*We* couldn't hand it over," says Mr. Bear. "PC Growler must do that since it's lost property. But you're right. It *is* odd they should try to break in to get what's theirs. Since we're responsible for it I shall keep watch tonight in case they try again." It is several hours later that Rupert is wakened by sounds of struggling and scuffling outside.

RUPERT SEES THE RAILIES CAUGHT

The Railies! But this time they fight
With Imps of Spring, who held them tight.

"Enough!" a voice calls, "Stop this row!"
And torchlight floods the garden now.

It's Growler who, with Mr. Bear,
Has caught the three Imps unaware.

"I'm going to sort this row out, see!
Come to the station now with me."

Rupert scrambles to the window. His father's feeling that the Railies might try for their train again has been proved right. What look like the same lot as before are out there. But so is a group of Imps of Spring and the Railies are struggling in their grip! Suddenly a voice rings out: "Enough! Stop this row!" And in the same instant the struggling group is caught in twin beams of light from torches held by Mr. Bear and PC Growler.

Rupert hurries outside. "You were right about keeping watch," he greets Mr. Bear. "He was that!" agrees PC Growler. "Your Daddy told me earlier about the attempt to break into your shed so I thought I'd look round here when I got back from Nutchester. I take it those tiny ones are the same lot. But who are the others holding them?" When he's been told who the Imps of Spring are he says, "Right, then. I'm taking you lot to the police station to sort this out."

RUPERT SEES THE IMP KING APPEAR

But how to get the Railies there?
"I've just the thing!" says Mr. Bear.

"This old birdcage will be just right,
We'll pop them in and lock it tight!"

"We'll have to take the Imps there too,
Perhaps my wheelbarrow will do!"

The Imps protest, and as they do
Their King steps forward into view.

The problem is how to get them all to the police station. Plainly the Imps won't be able to carry the Railies all that way with them struggling as they are. At the same time the Imps won't put them down in case they escape. It's Mr. Bear who has the answer. He produces a key from his pocket and hands it to Rupert with a whispered instruction. Rupert lets himself into the shed and re-appears with a birdcage. "We'll put the Railies in this," says Mr. Bear.

The Imps are delighted to see the Railies put into the birdcage. But their delight turns to rage when PC Growler asks Mr. Bear how they might take the Imps themselves to the police station and Mr. Bear suggests his wheelbarrow. "You can't take *us* there!" they squeal. "Oh, yes, I can!" Growler declares. "Our King will have something to say about this!" threatens one of the Imps. "I certainly shall!" a voice rings out and from behind a bush steps the King himself!

"I'm afraid," Growler starts to say,
"Your Imps were part of an affray!"

"Nonsense!" the King cries. "That can't be.
My Imps were just obeying me!"

"The Railies are the ones to blame!
Our home's been ruined since they came!"

The King's told he can't break the law.
The Railies hear this and guffaw.

When Rupert has introduced the newcomer Growler addresses him sternly: "Look, your Majesty, your Imps were involved in an affray." "Nonsense! They were obeying me!" the King retorts. "When I learned where their train was I knew the Railies would try for it again. I ordered an ambush and came along to watch. You must understand that we are entitled to punish these wretched creatures. Rupert Bear has seen what they have done to our home."

By the time the King has described the holed and sooty walls, with Rupert nodding agreement, it's plain (to Rupert, anyway) that Growler now sympathises with the Imps and he seems sorry to have to say, "I know how you must feel. But even though I don't think I can do anything about them, *you* can't just set on them here. You must understand it is against the law. And it's my job to see the law's kept." By now the Railies are scoffing and laughing at the King.

74

RUPERT MAKES A SUGGESTION

"Just promise your Imps will obey
The law and you'll be on your way."

A case like this needs an umpire
Thinks Rupert, and suggests the Squire.

"The Squire!" the Railies cry. "Oh, no!
Give us our train and let us go!"

The Squire's name seems to scare them all.
Says Growler, "Let's give him a call."

The Railies' laughter annoys Growler and he orders them to be quiet. Then he says to the King, "Look, just promise that your Imps will obey the law and you can all be on your way." "Oh, very well," sighs the King. "But what about the Railies?" Growler turns to them: "Yes, what about you? You may own the train but you're a great nuisance with it!" "Why not ask the Squire?" suggests Rupert. "After all he is a magistrate and should know what to do."

At once the Railies' delight at seeing the Imps told they're in the wrong disappears and is replaced by squeals of protest: "It's nothing to do with the Squire . . . Give us back our train and let us go!" But Growler has other ideas. He is very interested in the effect Rupert's mention of the Squire has had on them. "Rupert," he says, "that strikes me as a splendid idea. Let's all go down to the police station and I'll give him a call." The Railies wail.

RUPERT FOLLOWS GROWLER

Then with the train and little men
The barrow's loaded up again.

Perhaps it's just as well it's night,
This odd procession's such a sight!

The King comes to the station too.
"This way!" says Growler, "After you!"

"Good evening, sir. It's Growler here.
We need your help down here, I fear."

Like Growler the King of the Imps is struck by the Railies' being so upset at mention of the Squire and he says he'll come along with the rest of them to see what happens. He sends home two of his Imps, keeping one as an attendant. Soon as odd a procession as Nutwood has seen is making its way to the police station. The Railies in the birdcage ride with their train in the barrow pushed by Growler. Mr. Bear carries Rupert and the Imp King strides alongside. It's as well

this is happening in the middle of the night. Goodness knows what ordinary Nutwooders would make of a small King being ushered very respectfully into the police station by their local bobby while Rupert Bear carries a birdcage full of tiny people and Mr. Bear totes a little train. Inside, birdcage and train are ranged on the counter by Mr. Bear while PC Growler telephones the Squire to say he's needed and would he be good enough to join them.

Although the hour's so very late
Our lot do not have long to wait.

He sees the train and cries, "Bless me!
I don't believe that this can be!"

"When I was just a little boy
This train set was my pride and joy!"

Then one day he came down to find
It gone! And no trace left behind.

The Squire it turns out was in bed when Growler telephoned. But, as he says, when duty's to be done he'll do it. And when he's told that "people" must be locked up all night if he can't come and decide what should be done, he says, "I shall be there." And so he is, coat over his pyjamas and in his slippers. He is just asking what it's all about when he spies the little train. "Bless me!" he gasps. "I can't believe this! I never expected to see this again!"

"Do you know this train, sir?" Rupert asks him. "Know it!" echoes the Squire. "Why, that train was my pride and joy as a small boy." A dreamy look comes over his face as he remembers: "In every way it is a real little train. My father, a clever model engineer, built it. Its track took up an entire lawn. You should have seen it puff and smoke!" The Imp King mutters, "I *have*!" "Then one day it simply disappeared," the Squire says. "Track and all."

RUPERT MAKES A PLEA

"So you're the ones who stole my train!"
The Squire is very cross it's plain.

The King cries, when he's heard the tale,
"Constable, clap the lot in jail!"

"Oh please!" says Rupert, "Ask them why
They took the train. Let them reply . . ."

The Railies are released so they
May in all fairness have their say.

Now Rupert sees why the Railies were so upset at mention of the Squire. The little train, cause of all the trouble, isn't theirs, but the Squire's. The Squire is a kindly man but as he listens to Growler unfold the story of the Railies and "their" train his face darkens. "So it was you who stole my train!" he thunders at the Railies. "Clap them in jail, constable!" cries the Imp King, bouncing up and down and brandishing his sceptre at the little creatures.

"Oh, don't put them in jail!" At Rupert's cry the Squire and Growler turn to him. The Squire sinks to one knee and asks quietly, "But why not?" After all, they stole my train." Rupert who hates to think of anyone being locked up, says, "Maybe they were desperate. Oh, do ask them why they did it! At least let them have their say." "That seems fair," the Squire agrees. So the Railies are let out and one of them is chosen to tell their side of things.

RUPERT HEARS THE RAILIES' TALE

To see the world, the Railies say
From Lilliput they stowed away.

Because they were so small they found
They had to live beneath the ground!

For many years they lived this way
Until they saw the train one day.

It seemed to be the perfect home
For little folk who have to roam!

This is the Railies' tale: "When we were young we read of Gulliver's travels to our country, Lilliput and then we decided to have a look at *his* country. We stowed away on a ship coming here. At first it was exciting – flowers as big as frying pans. But it got too exciting. The cats were as big as elephants and thought we were some sort of mice. So we went underground where because we'd no home we had to keep moving, a wearisome business on foot."

For years (the tale goes on) the Railies trudged about underground, seldom venturing out. But one day they did. And there was what they'd dreamed of – a moving home, the little train and its tracks. The temptation was too much. So they took the lot underground, learned how to work it and became the Railies. "It's a bit late," ends the spokesman, "but – sorry!" While he has been talking an idea has been forming in Rupert's mind.

RUPERT HAS AN IDEA

Rupert's impressed by what he's heard
And asks if he may have a word.

"I'll run the train for charity
And hire the Railies to help me."

The Railies cheer. "How kind of you!"
But first there's something they must do . . .

They tell the quite astonished King
They'll come and clean up everything.

"Please, may I speak to you privately?" Rupert's request surprises the Squire. But he nods and takes him aside. As he listens to Rupert he nods and smiles. Then he turns to the others: "I like the idea Rupert has come up with. I have long meant to open the grounds of my home to the public for charity. But I have lacked an attraction. The train and tracks are going to be that attraction. And the Railies are going to help me. That should bring people in!"

How the Railies cheer! Then their spokesman addresses the Squire: "We don't deserve such kindness but before we join you there's something we must do." He turns to the Imp King: "We must clean up the mess we made in your place, sire." The King stares, grins and says, "If you can find a way to work that train without mess I can find a winter job for you – moving all our bulbs and seeds into position underground ready for our spring work."

RUPERT SEES HIS IDEA WORK

It's all arranged as Rupert said,
But now it's time he was in bed!

"I think I'll tell the Squire that he
Should install electricity!"

Because of all that Rupert's done
He's taken to the train's first run.

And when they see that Rupert's here
The Railies raise a rousing cheer.

It's all arranged. The Railies will clean the Imps' home, bringing out track for the Squire to have collected. Then they'll set it up in his grounds. But for Rupert it's back to bed and Mr. Bear wheels him home. Next morning Rupert recalls the Imp King's offer of winter work underground for the Railies if they can run the little train without mess and he announces, "I'm going to suggest the Squire has the train altered so that it can be run by electricity or stream."

One day some time later the Squire calls to take Rupert to the first appearance of the little train in his grounds. On the way he says, "I took your tip and had it fixed to go by either steam or electricity so that the Railies can use it underground for the Imps of Spring." When the pair of them reach the Squire's place the train is already whizzing round the track. The Railies aboard it raise a cheer when they see Rupert standing there. The End.

Whose Hat?

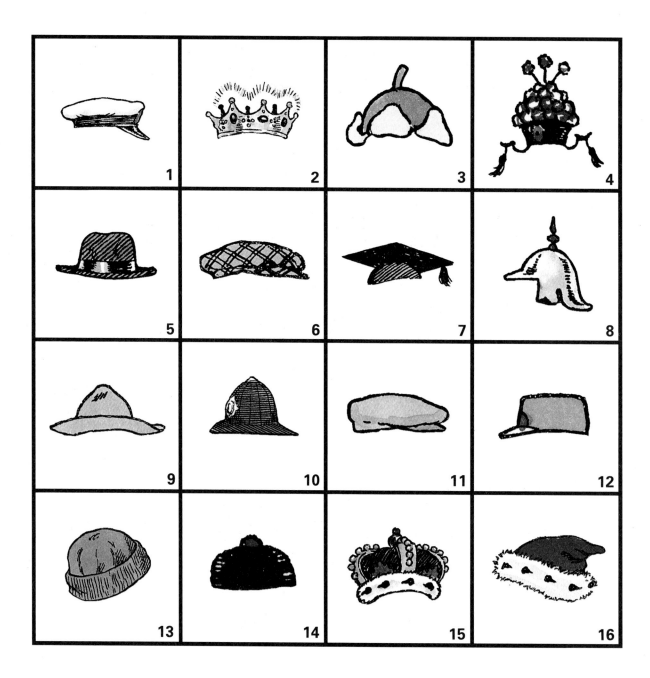

Each of the hats shown above belongs to a character you have met in this year's Rupert Annual.
Can you work out who owns each one?

(Answers on page 99)

RUPERT
and
Willie's Present

John Harrold.

RUPERT IS ASKED TO HELP

One morning the Bears think they hear
A small plane landing somewhere near.

When Rupert runs to where it lands
There Santa's little Cowboy stands.

He has a note for Willie Mouse.
Could Rupert take it to his house?

"Santa don't like his folk seen, so
That's why I'm asking you to go."

One morning not long before Christmas the Bear family are at breakfast when Mr. Bear says, "Listen! Isn't that a small aeroplane I hear?" "It is," agrees Rupert. "What's more it sounds as if it's landing near here. I must go and look after breakfast." So not many minutes later he is hurrying towards where he thinks the aircraft has landed. And, yes, there it is. So is its pilot who turns out to be someone Rupert knows quite well – the little Cowboy who is Santa

Claus's flying helper. "Hey, Rupert!" he calls. "Can you spare some time to do me a favour?" He produces a letter. "This here's for your pal Willie Mouse. Will you deliver it for me? You know Santa don't like his people to be seen if we can help it." Now, Rupert knows this is so, but he can't help feeling that the Cowboy is unusually keen for someone else to deliver the letter. But he gladly agrees and promises to return and confirm that Willie has got it.

RUPERT'S PAL IS UPSET

As Rupert hurries on his way
He wonders what the note can say.

At Rupert's knock his pal appears.
His face lights up at what he hears.

"For me? From Santa?" Willie cries.
But as he reads tears fill his eyes.

"What's wrong?" asks Rupert, mystified,
As Willie turns and runs inside.

So Rupert sets off for Willie's cottage and as he hurries along he wonders why the Cowboy was so keen not to take the letter himself. It can't be because he's nervous about Willie. The little mouse is the most timid soul in Nutwood. Everyone likes him but the fact remains he *is* awfully timid. Rupert is still puzzling as he knocks on Willie's door. The door opens a little and Willie peeps out. "A letter from Santa for you," Rupert announces breezily.

"For me? From Santa?" The timid little mouse squeaks excitedly and his face lights up. He rips open the envelope and starts to read the note it contains. But as he reads his expectant smile is replaced by a look of disbelief. Then while Rupert looks on aghast he gulps and bursts into tears. "Oh, Willie, what's wrong?" Rupert cries. "Can I help?" His answer is a wail from Willie who crumples the letter and bolts back into his house, slamming the door behind him.

RUPERT LEARNS WHAT'S WRONG

He goes to let the Cowboy know
The note's upset poor Willie so.

Willie's asked Santa for a thing,
It seems, that Santa cannot bring.

"If Santa told me what, then I
Could see Willie and tell him why."

"I'll ask my mother if I may,
Then come to join you straightaway."

What on earth's wrong! Rupert stares at Willie's door and wonders what to do. He knocks. "Willie, what's wrong?" he pleads. "Maybe I can help." But the door stays shut and from inside comes only the sound of Willie's sobs. At last Rupert gives up and goes back to the Cowboy to tell him the effect on Willie of the letter from Santa. But the Cowboy doesn't seem at all surprised. "Aw, gee!" he groans. "I thought this might happen. Sorry, Rupert!"

"You expected this?" gasps Rupert. "Yup," the Cowboy admits glumly. "It was wrong of me not to take it myself. It said Santa can't give Willie what he asked for – but don't ask what *that* is." "If I did know I might be able to help him," Rupert argues. "Only Santa could tell you *that*!" protests the Cowboy. "Then let's go and ask him!" Rupert pleads. The Cowboy pauses, then, "OK!" he says. And next thing Rupert is dashing home to ask his mother if he may go.

RUPERT FLIES TO SEE SANTA

"Of course!" she answers. "I agree
But do be back in time for tea!"

Keen to be off, he hurries straight
To where the Cowboy said he'd wait.

The plane takes off and up they fly
To Santa's castle in the sky.

They land. The Cowboy leads the way
To hear what Santa has to say.

Mrs. Bear thinks Rupert is joking when he runs in pleading, "Please, may I go to see Santa Claus about something very important?" But when she hears the whole story she says, "Poor Willie. Such a nice little person and so timid! Yes, of course, you must help him. So off you go to see Santa, but be back in time for tea." And so a few minutes later the Cowboy, waiting in his aircraft, sees Rupert racing up shouting, "It's all right! Mummy says I may go!"

The Cowboy promises to get Rupert back in time for tea, Rupert climbs aboard and in no time at all the pair are high in the clear sky. As they fly, Rupert keeps wondering what it can be Willie has asked for that Santa can't bring him. It's not as if Willie's greedy. At last, rising from a bank of cloud, Santa's castle appears and soon the Cowboy has landed his machine on a wide terrace guarded by toy soldiers and is leading the way at a run to Santa's office.

RUPERT IS TOLD ABOUT WILLIE

The Cowboy says that Rupert's come
Because he wants to help his chum.

"What Willie asked for was to be
As brave as all his pals, you see."

Rupert agrees not to betray
A word of what he's learned today.

"If you can help your pal do try!"
Cries Santa as he waves goodbye.

Rupert has met Santa several times but he has never seen him look as stern as when the Cowboy reveals that Rupert knows that he has had to refuse Willie the present he wants. But his look softens when he learns that Rupert does not know *what* Willie asked for, that Willie is so upset and why Rupert has come to see him. "Oh, dear me, poor little chap!" he says. "And you, Rupert, think you may be able to help him if I tell you what he asked for? Well, I'll tell you, but you must promise to tell no one what I say." Rupert promises and Santa continues, "Willie asked to be as brave as his pals. That's something *I* can't give him." "But we're not at all brave!" protests Rupert. Santa sighs: "But what counts is that Willie believes you are. Do you still think you can help him?" "I'll have a good try," Rupert says. But now it's time to get back to Nutwood. Santa follows the pair outside. "Do try to help him, Rupert!" he calls.

RUPERT DECIDES ON A PLAN

"Somehow I must make Willie see
We're really no more brave than he!"

Already by the time they land
The small bear has a scheme all planned.

Then, as he hurries home for tea,
He sees his pals, "The very three!"

The four of them agree that they
Will meet in Rupert's shed next day.

On the way back to Nutwood Rupert wonders how he is going to help Willie. "He believes we're brave," Rupert reasons. "We're not. We're just not as timid as he is. Now, if we could get him to believe that we *are* as timid . . ." And he keeps turning the thought over in his mind so that by the time the Cowboy lands on a quiet part of Nutwood Common Rupert has a plan. "You look a lot more cheeful," says the Cowboy. "Yes, I think I know what to do," Rupert tells him.

The Cowboy leaves and Rupert starts for home. He's almost there when he sees Algy Pug, Edward Trunk and Bill Badger. "Just who my plan needs," he thinks. "But I can't tell them about Santa and what Willie wants." So when the three want to know where he has been he answers them with a question: "Will you come to a secret meeting tomorrow?" "Ooh, yes!" they chorus. "Right," says Rupert. "My garden shed after breakfast. See you then." And off he hurries.

RUPERT GETS HIS PALS TO HELP

"I came," says Bill. "But, as you see,
My brother had to come with me!"

"Willie won't feel so bad if we
Make him think we're as scared as he."

The pals ask Willie if he'd like
To join them on a river hike.

Willie agrees and soon they find
The path that Rupert has in mind.

Because his mother is busy Bill has to bring his little brother to next day's meeting. But Baby Badger is a happy little soul so no one minds. When everyone's settled Rupert begins:"I can't tell you *how* I know, but Willie Mouse is very upset because he's so timid. I think he'd feel better if he believed *we* were just as timid." "But we're not!" protests Algy. "Well, we're jolly well going to make him think so," Rupert says. And he tells the others his plan.

When he's quite sure they understand it, Rupert leads the way to Willie's house. All four try to look as usual as they can when Willie comes to the door. "We're going for an explore along the river and we thought you might like to come," Rupert says. "Are you sure?" Willie asks. "Of course!" chorus the others and Baby Badger who can't talk, chuckles. So a few minutes later Willie is marching down to the river with his chums to start their "explore".

RUPERT PREVENTS A BLUNDER

"It's like a bridge, that big branch there.
Let's cross the river – if you dare!"

"That's easy!" Algy scoffs. But no!
It seems it's too far – he won't go!

"You're right!" The other three agree.
They wouldn't dare cross by that tree!

They carry on till Rupert sees
A way to cross the stream with ease.

Willie is quiet as Rupert leads the way up-river. The others laugh and joke, pretending they don't notice. They reach a spot where the river races between steep banks. A stout branch spans it like a bridge – a simple crossing for anyone young and nimble. "Who dares cross it?" cries Rupert. "Too easy!" Algy scoffs, starting towards the tree. "Ahem!" Rupert coughs loudly. Algy stops. For a moment he forgot the plan. "Too easy to, er, fall off," he says lamely.

So the pals leave the tree-bridge and carry on up the river. As they go, with Baby Badger gurgling happily in his push-chair, Rupert, Bill, Edward and Algy start to talk loudly about how foolish and dangerous it would have been to cross the branch. Willie tramps on in silence, speaking only when spoken to and not saying much then. After a little Rupert stops again. "Ah, here's another way across," he says and points to a line of large easy-looking stepping-stones.

RUPERT FINDS A ROPE SWING

"Come on!" cries Edward. "Race you there!"
"Come back!" calls Rupert. "Take more care!"

Eventually the pals decide
They won't cross to the other side.

The chums continue on their way.
They don't seem very brave today!

Then acting as if in surprise,
"Oh, look! A rope swing!" Rupert cries.

"Race you across!" At Edward's shout Rupert swings round to see his chum thundering down the bank. "Silly ass!" he thinks. "He's forgotten for the moment, too!" As Edward lumbers past Rupert cries, "Edward, be careful!" Edward skids to a halt. "Oh, dear, I forgot –," he begins, then stops and says, "You're right it does look unsafe." The others make a big show of agreeing with him. "Don't you think so, Willie?" they ask. "I suppose so," he answers dully.

As the pals press on again Baby Badger is the only jolly one of the party, and he's too little to understand what's being said. Willie walks in silence not joining in the talk of Rupert and the others about how you can't be too careful and should never take risks. Now, Rupert knows this stretch of the river well and he has often crossed it by a rope swing which is where he is now headed. But he pretends surprise when it appears. "Look! A rope swing!" he cries.

RUPERT'S PAL SAYS HE KNOWS

He takes the rope, but then, "Oh dear!
It's much too dangerous, I fear!"

"It's not!" a voice comes from behind,
"You know it's nothing of the kind!"

"You're pretending!" he starts to say
But then the pals gasp in dismay.

The push-chair has begun to slide
And speeds towards the riverside.

"I must have a go at this," Rupert says as he frees the end of the rope which has been tied back. The rope is stout, the branch it hangs from is thick and the river narrow – an easy swing. But instead of launching himself across Rupert hesitates. "N-no," he quavers at last. "It's far too dangerous, I fear." "No, you don't, Rupert! You don't think anything of the kind!" Everyone turns to stare at the speaker – Willie. The silence is broken only by the chuckling of Baby Badger and the squeak of his push-chair as he bounces happily. Then Willie goes on glumly: "I can see you're all pretending to be timid like me so that I won't feel left out." "Oh, no!" the others chorus. "But yes!" Willie insists. "Oh, no!" the others repeat. But now there is alarm in their cry and they are looking beyond Willie. Willie turns and he squeals, "Oh, no!" For Baby Badger's bouncing has freed his push-chair brake and it is plunging towards the river!

RUPERT JOINS A RESCUE BID

The others freeze. They move too late.
But Willie doesn't hesitate.

He plunges in without a thought
And, just in time, the push-chair's caught.

To cling on tight's all he can do.
"Hold on!" yells Rupert, "I'll help you!"

"Get ready, when I give a shout,
To pull my scarf and haul us out!"

Rupert, Algy, Bill and Edward are frozen to the spot in horror as Baby Badger's push-chair careers towards the river. It is Willie – the so timid little mouse – who acts. He does not hesitate before hurtling after the push-chair to hit the racing water at almost the moment it does. He gasps at the icy cold but plunges after the little badger who is being swept rapidly downstream. The others race to the water's edge as Willie grabs for the push-chair. For one terrible moment it seems Willie will be dragged along by the push-chair he has managed to catch hold of. But with his free hand he grabs a rock and clings grimly to it. "Hang on!" Rupert cries. "I'm coming in to help you." He tears off his scarf and, clasping one end, pushes the other into Bill's hands. "Take this and when I shout you lot get ready to pull us out," he tells the others and then braces himself to jump into the icy water.

RUPERT HELPS HIS LITTLE PAL

Then, clutching the scarf's other end,
Rupert jumps in to help his friend.

The river is so fast the pair
Can do no more than hold the chair.

"Quick!" Edward cries. "Hold this branch tight!"
Then starts to pull with all his might.

The other pals know what to do,
They brace themselves and tug hard too.

Bill, clutching one end of Rupert's scarf, sets his feet firmly on the bank. Algy clasps him round the waist to help take the strain. "Here goes!" cries Rupert and jumps into the river. The cold takes his breath away but he lunges at the push-chair and grabs the handle. As he feels the pull of the racing river he can't imagine how Willie has held on so long. Now Edward climbs onto a rock on the bank with a long branch clutched in his trunk. He swings it over the water to Willie. "Hold this tight," he mumbles, for you can't speak clearly with your trunk round a branch. Willie lets go of the rock and grabs the branch. When Rupert is sure they are as ready as they're going to be, he cries, "Haul away!" The three on the bank heave as hard as they can. For what seems a very long moment nothing happens. "Harder, harder!" Rupert urges. Then slowly, steadily, Baby Badger and his rescuers are pulled to safety.

95

RUPERT'S LOT HURRY HOME

As Baby Badger's lifted clear
The pals, though breathless, raise a cheer.

To warm themselves the pals now need
To dash to Bill's house at top speed.

Bill's mother gasps, then, straightaway,
Whisks them inside without delay.

While Baby's bathed the other pair
Are both brought dressing gowns to wear.

The moment the push-chair is close enough, the three on the bank pull it clear of the water then help Rupert and Willie to dry land. Breathless as they are, they manage to raise a cheer. Even Baby Badger gives a shivery chuckle. At once Bill peels off his jacket and wraps it round his little brother. Algy gives Rupert his, and Willie whose teeth are starting to chatter, is tied into Edward's coat with Rupert's scarf. Then they start at a trot for Bill's house.

Mrs. Badger gives a squeak of alarm when her sons and their chums turn up on her doorstep. They are whisked inside before they can start to explain. A warm bath is put before the living-room fire and Baby Badger is undressed and popped into it, chuckling at being the centre of so much attention. While this is happening Bill is sent to fetch dressing gowns and pyjamas for Rupert and Willie. "Now, into those things, you two," commands Mrs. Badger.

RUPERT SAYS WILLIE IS A HERO

Then, over tea, Bill tells the way
His pals' quick thinking saved the day.

Bill's mother praises all they've done,
But's told that, "Willie's the brave one!"

Then, at Rupert's request, she brings
The little mouse some writing things.

"Dear Santa Claus, I've changed my mind.
I've got what I asked for, I find."

When everyone is warm and dry again and settled down to milk, sandwiches and cake, it's the time for explaining. Out comes the story of the "explore" that went wrong. No one, though, mentions Rupert's plan that led to it for that would embarrass Willie. At last Mrs. Badger rises and embraces the two heroes. "I have never heard of anything so brave!" she cries. "Oh, let's be clear! Willie here was the really brave one," Rupert says. Willie blushes.

"How can I thank you?" Mrs. Badger asks. The others stare when Rupert says, "By letting Willie have writing things, please." As a puzzled Mrs. Badger fetches them from a desk Rupert adds, "He should really write to Santa Claus. There's not much time left before Christmas." Willie blinks. Then he grins, picks up a pencil and he writes, "Dear Santa Claus, I have changed my mind. I find that, after all, I've got what I asked for before." THE END.

97

Rupert's Memory Test

Please don't try this test until you have read all the stories in the book. When you have read them, study the pictures below. Each is part of a bigger picture you will have seen in a story. When you have done that, see if you can answer the questions at the bottom of the page. Afterwards check the stories to discover if you were right.

CAN YOU REMEMBER . . .

1. What is the name of this Emperor?
2. Who are these men?
3. What does this control?
4. Who is this?
5. What are Rupert and his father doing?
6. Who lives here?
7. What is the name of this ship?
8. Whose door is this?

9. Where are these mountains?
10. What is this?
11. What has Rupert found?
12. Who is Growler talking to?
13. Where is Bill arriving?
14. Who is Willie writing to?
15. What is the name of this island?
16. What lies behind these gates?

Follow
Rupert

each day in the
Daily Express

John Harold.

ANSWERS TO PUZZLES

Spot the Difference: (page 58) 1. Lamp missing; 2. Bobble missing Conjurer's hat; 3. Conjurer's pigtail gone; 4. Conjurer's belt buckle missing; 5. No checks on Rupert's trousers; 6. Tigerlily's satchel gone; 7. Flower missing Tigerlily's dress; 8. Sage of Um's spectacles missing; 9. Saucer missing from table; 10. Spout missing from coffee pot; 11. Handle missing from coffee cup.

Whose Hat? (page 82) 1. The Customs Officer; 2. Figurehead of the 'Sea Queen'; 3. An Imp of Spring; 4. The Emperor Chan; 5. Wing or Wang; 6. Scrogg; 7. Dr. Chimp; 8. One of Chan's soldiers; 9. The Little Cowboy; 10. PC Growler; 11. The Squire; 12. One of the Railies; 13. Sailor from the 'Sea Queen'; 14. The Chinese Conjurer; 15. The King of the Imps of Spring; 16. Santa Claus.

© 1990 EXPRESS NEWSPAPERS p.l.c.   Printed in Great Britain by Jarrold Printing, Norwich